Iness Jabri

Food hygiene and safety

Iness Jabri

Food hygiene and safety

The 5 M's and the HACCP method

ScienciaScripts

Imprint

Any brand names and product names mentioned in this book are subject to trademark, brand or patent protection and are trademarks or registered trademarks of their respective holders. The use of brand names, product names, common names, trade names, product descriptions etc. even without a particular marking in this work is in no way to be construed to mean that such names may be regarded as unrestricted in respect of trademark and brand protection legislation and could thus be used by anyone.

Cover image: www.ingimage.com

This book is a translation from the original published under ISBN 978-620-2-27815-7.

Publisher:
Sciencia Scripts
is a trademark of
Dodo Books Indian Ocean Ltd. and OmniScriptum S.R.L publishing group

120 High Road, East Finchley, London, N2 9ED, United Kingdom
Str. Armeneasca 28/1, office 1, Chisinau MD-2012, Republic of Moldova, Europe
Printed at: see last page
ISBN: 978-620-5-88116-3

Table of contents :

Introduction

Definition:

Hygiene is the set of measures to be respected in order to preserve health. In IAA, Food Industry, these measures allow the obtaining of healthy food.

Food hygiene has two components (*Codex Alimentarius*):

> **Safety** = food without dangers (no salmonella, no glass bits...)

> **Safety** = acceptable, edible food (no bad odor, no contamination)

Importance of hygiene:

Obvious sanitary importance: less hygiene = more illnesses.

Economic importance: longer shelf life of the product, possible exports, accidents avoided. A hygiene mistake is often the death of the company.
Regulatory significance.

Hygiene is necessary from the moment the product is received until it reaches the consumer's plate.

The main principles of hygiene:

If we represent on a cause-effect diagram the problem of lack of hygiene we find that the major causes of "non-hygiene" = the "5M" Materials, Material (equipment), Environment (premises), Method and Manpower.

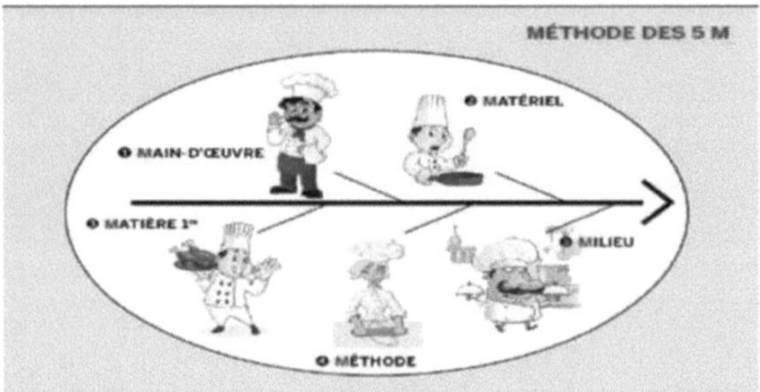

Chapter 1: Establishment (Design and Facilities)

1. Location:

1.1. Establishments:

When deciding where to locate food production facilities, consideration should be given to potential sources of contamination, as well as the effectiveness of any reasonable measures that could be taken to protect food.

No establishment should be located where, after considering such safeguards, it is clear that a threat of contamination will remain to the safety or suitability of the food.

In particular, facilities should be located at a great distance:
- Flood prone areas, unless adequate safety features are in place.

- Polluted areas and industrial activities that represent a serious threat of food contamination. (the distant environment contaminates: the factory or workshop must be far from the sources of contamination *(e.g.: minimum distance from a road =5m, a house =50m, a farm =100m, a waste stock =200m).*
- Areas subject to pest infestations.

- Areas where waste, solid or liquid, cannot be effectively removed.

1.2. Materials:

Equipment includes machines, tools, tables, conveyors, bins, etc. The equipment will be cleaned and disinfected often, and many machines are subject to Cleaning In Place (CIP) (see next part of this course). This requires proper design and materials:

> The equipment must be suitable for the activity, of simple design, without sharp angles or blind spots, and easily removable.

> The materials in contact with the food (= food surfaces), must be

compatible with the food, waterproof, resistant, smooth and rot-proof.

1.3. Premises and rooms :SMALADE

S- **Separate sectors**. Incompatible" sectors must be physically separated. It must not be possible to go directly from a soiled area to a healthy area (*e.g.: reception/manufacturing*). Same thing between hot and cold areas (otherwise, water condensation on the ceiling, sources of bacterial "rain").

M- **Forward motion** is imperative. The circuit of the products must not include **any backward movement or crossing**: We go from dirty to clean, to avoid cross-contamination (e.g.: *the materials I do not cross the treated product; the food does not cross the cleaning products*).

Separate sectors and forward motion can be seen on the plant layout

A- **Rational layouts**: easy to clean "shapes" (*no corners or* Floor slope > 1%, floor drainage with grated siphon.) No sharp corners, but rounded grooves (wall-ground). No dust nests (*e.g. cables, shelves*)

Wide space between wall and equipment, and around each machine. Machines on sealed feet.

Lighting without corners (dust, flies) and sufficient (*e.g.: 220 lux in the manufacturing room*)

L- **Washable Materials** (floors, walls, ceilings, doors & windows): waterproof, resistant, smooth, clear but non-slip (floors) (*e.g. tiles, resin or vitrified cement*). Beware of joints!

A - **Controlled air**. Two complementary aspects:

- **renew the indoor air** to eliminate endogenous contamination (fumes, smoke, aerosols, food particles, human desquamation).
- **filter the outside air** to eliminate dust and bacteria. This requires a central
- ventilation/filtration, filter control, air flow control (overpressure area). To protect ultra-sensitive areas, we are starting to use sterile air flows, in

close protection.

D- **Waste**: major source of direct and indirect contamination (pests). Closed and watertight garbage cans in a specific closed room. Particular type of waste: effluents (generally liquid). The law imposes to control them to limit pollution.

E- **Drinking water**. (*decree of 3 Jan. 89*) Often the need for water is enormous, leading the food industry to have its own borehole (*e.g.: 10,000 m3/d for a poultry slaughterhouse*). This obliges to treat the water and to control its potability (bacteriological analysis) and hardness (scale). Other fluids: provide enough steam, cold, compressed air. Electric cables and pipes = nooks and crannies.

2. Design and layout

It is within this framework that the principle of the forward march (respect for the forward march) is integrated.

2.1. The march forward

<u>Definition</u>: The principle of forward motion consists in organizing the flows of all the participants according to the sensitivity of the product and the risks of contamination inherent in the industrial process.

<u>Speakers:</u>

-Products: raw materials, ingredients, packaging, work in progress, finished products, waste...

-Materials: clean materials, dirty materials, mobile materials (pallets, pallet trucks, carts)

-The staff: of the production workshops, maintenance visitors, visitors

<u>Objectives:</u>

Limit the risk of cross-contamination between contaminated and fragile foodstuffs

<u>Means:</u>

-Physical separation of clean and soiled areas

-Separation in time

-Circuits without return or crossing

The design of the premises must be thought out so that at no time do the different circuits of the workers overlap to cause the appearance of cross contamination.

If the spatial arrangement of the premises does not guarantee the absence of crossings, then a forward march in time must be applied. The operations are carried out at different times. This time lag often requires cleaning and disinfection between operations. If possible, start with the least contaminating operations

2.2. Organization of flows:

MP- PF

As far as the products are concerned, it is a question of going from the raw material to the finished product without ever going back. Generally speaking, there should be no cross-fertilization between products at different stages of the manufacturing process.

Separations are also recommended for the following areas:

- Separation of cold areas (pastry) / hot areas (kitchen): condensation phenomenon
- Separation of soiled areas (receiving dock, garbage room), inert areas (kitchen) / sensitive / ultra-sensitive areas (chopping room, grinding room)
- Separation of dry areas (commissary) / wet areas (cooking room): the ambient humidity is a factor favorable to the surface microbial growth.

The material

The principle of forward motion also applies to equipment, i.e., equipment should not be moved to "cleaner" premises after having been in "less clean" premises without the application of appropriate measures (cleaning and disinfection).

A simple, linear arrangement of processing equipment is preferred. Movement of mobile equipment (e.g. pallets, carts, etc.) from one room to another should be limited and controlled to minimize the risk of cross-contamination.

The Staff

The same principle applies to personnel, who should always move from the area most susceptible to contamination (where the product is most fragile) to the area least susceptible (where the product is least fragile), but never vice versa. If the reverse must be done, measures must be taken (change of clothes, disinfection of shoes and hands, etc.).

Four main types of establishments can be defined according to the modalities of the forward march, namely the linear establishment, the U-shaped establishment (type 1 and type 2), the L-shaped establishment and the gravity establishment.

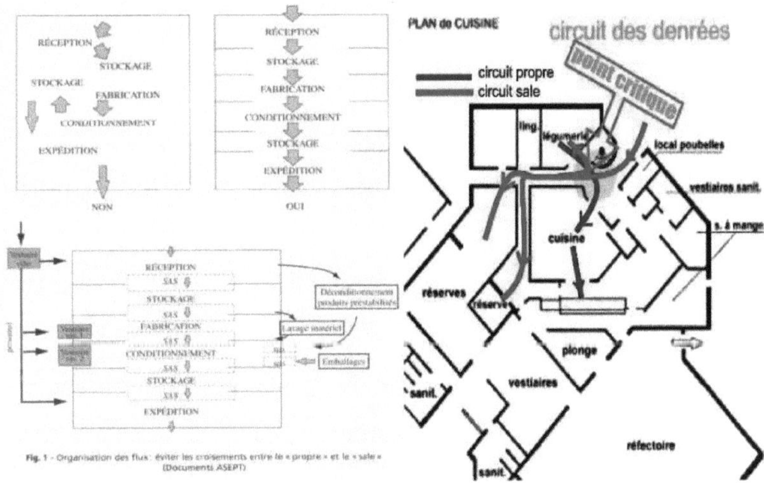

Organization of the MP-PF flows in a hotel kitchen

2.3. Structures and internal accessories:

Structures within food production establishments should be of solid construction and durable materials and should be easy to maintain, clean and, where necessary, disinfect. In particular, the following specific criteria should be met where necessary to maintain the safety and wholesomeness of food products

- ❖ The surfaces of walls, partitions and floors should be made of waterproof materials, walls and partitions should have a smooth surface up to an appropriate height,

- ❖ Floors should be constructed to allow for proper drainage and cleaning (presence of slope),

- ❖ Ceilings and ceiling fixtures should be constructed and finished to minimize dirt accumulation, vapor condensation, and peeling.

- ❖ Windows should be easy to clean, constructed to minimize dirt accumulation and, if necessary, have removable insect screens that can be cleaned. If necessary, windows should be sealed.

- ❖ Doors should have a smooth, non-absorbent surface and should be easy to clean and, if necessary, disinfect.

- ❖ The wall/floor/ceiling connections are to be made round and watertight,

- ❖ Work surfaces that come into direct contact with food should be in good condition, durable and easy to clean, maintain and disinfect. They should be constructed of smooth, non-absorbent materials and remain inert to food, detergents and disinfectants under normal working conditions.

Other points to consider:

- ❖ The exterior of the building shall be designed, constructed and maintained to prevent the entry of contaminants and unwanted animals. There should be no unprotected openings, properly located vents, and the roof, walls and foundation should be maintained to prevent leakage;

- ❖ Drainage and sewage systems shall be equipped with appropriate traps and vents;

- ❖ Facilities shall be designed and constructed so that there is no interconnection between the sewer system and any other effluent system in the facility;
- ❖ Sewage and effluent systems should not pass directly over or through production areas unless they are under control to avoid contamination;
- ❖ Paints, chemicals, lubricants and other materials used to treat surfaces or equipment that may come into contact with food must not contribute to unacceptable contamination of food.

2.4. Temporary / mobile locations and vending machines:

The premises and structures considered here are market stalls, street and mobile sales vehicles and temporary premises in which food is handled (e.g. tents). These premises and structures should be located, designed and constructed in such a way as to avoid, as far as possible, contamination of food products and entry of pests.

2.5. Materials:

General considerations:

Equipment and containers (other than non-reusable containers and packaging) that come into contact with the food should be designed and constructed to ensure, where necessary, that they can be adequately cleaned, disinfected and maintained to prevent contamination.

Equipment and containers should be made of materials that are non-toxic for their intended use. Where necessary, equipment should be durable and removable or capable of being disassembled to allow for maintenance, cleaning, disinfection, monitoring and to facilitate the detection of pests.

Heat treatment or cooling equipment:

Equipment used to cook, heat treat, cool, store or freeze food products should be designed so that the required temperatures are reached as quickly as necessary to ensure the safety and wholesomeness of the food and are maintained effectively.

It should also be designed to allow for temperature monitoring and adjustment (display, recorder, calibration, probe thermometer).

Where necessary, it should have an effective means of controlling and monitoring humidity, airflow and any other characteristics that may adversely affect safety or edibility.

These specifications are intended to ensure that:

J Harmful or undesirable microorganisms or their toxins are eliminated, or reduced to safe levels, or their survival and growth are effectively controlled.

J Where necessary, critical limits established in HACCP plans are monitored.

J Temperatures and other conditions necessary to ensure the safety and wholesomeness of food can be promptly achieved and maintained.

Containers for waste and inedible substances:

Containers for waste, by-products and inedible or hazardous substances should be specifically identifiable, properly constructed and, where necessary, made of impervious material. Containers used for hazardous substances should be identified and, where appropriate, lockable to prevent deliberate or accidental contamination of food products.

2.6. Facilities

Water supply:

An adequate supply of potable water, with appropriate facilities for storage (no corrosion, no contamination), distribution and temperature control, should be available whenever necessary to ensure the safety and wholesomeness of food products.

Drinking water should meet the microbiological criteria standardized in the latest edition of the World Health Organization's guidelines for drinking water quality, or be of superior quality:

   J Total germs < 100 / ml,

   J Total coliforms < 1 / 100ml,

   J Fecal coliforms < 1 / 100ml,

   J Fecal Streptococci < 1 / 100ml,

   J Clostriduim sulfito-reducing < 1 / 20ml.

Non-potable water (e.g., for fire fighting, steam generation, refrigeration and similar uses that do not contaminate food products) must be piped separately. Non-potable water lines must be identified and not have any connections or backflow into potable water lines.

Drainage and waste disposal:

Appropriate facilities should have adequate drainage and waste disposal systems and facilities. These should be designed and constructed to avoid the risk of food contamination.

Cleaning:

Appropriate and properly designed facilities should be provided for the cleaning of utensils and equipment that come into contact with food products. Where necessary, they should be supplied with hot and cold potable water.

Sanitary installations:

These facilities should be appropriately located and marked.

The locker room is a buffer zone, an obligatory passage between the outside environment, the dirty zone, and the company, clean area. The locker room must therefore be attached to the production premises.

Before entering a workshop, a module or a defined area, personnel must systematically go through a changing room to put on the appropriate clothing for their function or task.

In some cases, there may be several changing rooms. For example, in meat or poultry slaughterhouses, packing personnel have a separate locker room from those working on the actual slaughter. In other cases, company policy may encourage a central locker room. In this case, the personnel wears a minimum circulation uniform (called minimum mandatory uniform) which can be reinforced when accessing a risk area (e.g. packaging) through an airlock.

It is often through the design and the functioning of a checkroom that we can discover the organizational approach and the hygiene culture of the company. The state of the changing room (cleanliness, smell, visual aspect, etc.) is the mirror of the quality in the company and of the products or services provided. The design of the changing rooms is therefore of great importance to ensure hygiene and to achieve consistency in the control of contamination in the production workshops.

Temperature control:

Depending on the nature of the operations performed, there should be adequate facilities for heating, cooling, cooking, refrigerating and freezing food, for storing refrigerated or frozen food, and for monitoring its temperature and, if necessary, for controlling the ambient temperature to ensure the safety and wholesomeness of the food.

Air quality and ventilation:

Adequate natural or mechanical ventilation should be provided, in particular

for :

J Minimize airborne contamination of food products (e.g., aerosols and condensation water).

J Check the room temperature,

J Avoid odours that may affect edibility.

J Prevent moisture, if necessary, to ensure the safety and wholesomeness of the food.

For the past fifteen years, the concept of clean room has been developing in the food industry. It is a response to the need for increased product safety, but also to the need to extend their life span. These facilities are based on the control of air quality. This control concerns in particular the temperature, hygrometry and especially dust conditions and consequently the objectives in terms of aerobio-contamination.

The installations of the aeraulic system (air handling unit, circulation ducts and diffusion and return elements) must be designed in compliance with hygienic design criteria. The materials of the air handling unit (ducts) must be smooth and non-absorbent.

Connections inside the plant should be rounded off. As far as possible, the accessibility to the cleaning of these ducts is to be foreseen. As far as the elements such as the supply or return vents are concerned, they must be easily dismantled to ensure cleaning.

If screens are installed, the materials must be resistant to the cleaning products commonly used in the food industry (bases and acids). The design of these grids must not be able to constitute retention zones. During the design of the premises, the installation of airlocks to manage the flows must be foreseen.

Lighting:

Adequate natural or artificial lighting should be provided to enable the worker to operate in hygienic conditions. The intensity of the lighting should be appropriate to the nature of the operation. Lighting devices should, where necessary, be protected to prevent contamination of food in the event of breakage.

Storage:

Where necessary, adequate facilities should be provided for the storage of food, ingredients and non-food chemicals such as cleaning products, lubricants and fuels.

Where appropriate, food storage facilities should be designed and constructed to:

J Allow for proper maintenance and cleaning.

J Prevent access and establishment of pests.

J Enable effective protection of food from contamination during storage.

J Provide an environment where necessary to minimize spoilage of food products (e.g., through temperature and humidity control).

Chapter 2: Personnel Hygiene

Personnel are one of the main sources of microbial contamination of food. They can be the source of pathogenic microorganisms, either because they are sick, or because they are healthy carriers, or because they serve as simple vectors carrying germs from a contaminated surface to the food being handled, by the hand, gloves, clothing, or equipment used. Thus, all necessary standards must be implemented so that human contamination is minimized as much as possible.

The hygiene of the staff is its health, its body cleanliness, its dress and its motivation, reinforced by training and also by appropriate, well-designed and well-maintained facilities and equipment.

The staff is the weakest, most important "link" in hygiene control:

- It conditions the other "M": it controls the raw materials, it cleans the equipment, it implements the environment (e.g. separate sectors), it "makes" the methods.
- It is a major source of germs, both common (1011 bact/g stool) and pathogenic. The LPN staff must therefore be clean, healthy, trained/hygiene, and trained/working. (So neither Dirty, nor Sick, nor Ignorant of hygiene, nor Confused about techniques).

Body cleanliness:

- Equipment and procedures must enforce cleanliness.
- Sink, one per workshop, non-manual control, liquid soap, "sterile" hand towels (paper or roll). Obligation to wash hands frequently and carefully
- Clean, pedal-operated toilet with sink, separated by 2 doors from risk areas
- Footbaths at the entrance to risk areas, new disinfectant every day.
- Essential work clothes: depending on the level of risk.

    ♪ Hair cap (80000 particles >0.5^m/m3/min)

ᴊ Specific boots of the workshop, passed to the foot bath, dried the evening.

ᴊ Clear blouse, without pockets or buttons, in polyester or tergal.

Good health:

A sick person should be kept away from "risk" positions, as long as his illness increases the excretion of pathogens (diarrhea, cough, sore throat, fever, boil or panicitis).

Cuts and scratches on exposed skin must be covered with a bandage (blue, with detectable metal wire), provided by the organization, and protected by gloves.

Many people are healthy carriers of pathogens, especially in the food industry (*e.g. Salmonella: 10-25% of people, C. perfringens: 30-70%, Listeria m.: 515%*). However, the excretion of germs is discontinuous, random, and can last for years. In practice, it is necessary to act as if everyone was a healthy carrier.

Trainings :

The personnel must be trained and made aware of hygiene on a continuous basis (posters, instructions, supervision). Directive 93/43/EEC requires **a** professional training plan

No Confusion on techniques:

Procedures should be written and posted in addition to job training upon hire. Management should pay special attention to temporary workers, interns and substitutes.

The management is responsible for hygiene: Not only do managers and directors have to invest in hygiene (*Haccp, training, investments*), inspect the workshops often, but they also have to strictly respect the hygiene rules (*e.g.: do not no smoking, protective clothing, foot bath, especially with visitors).*

1. <u>Health status:</u>

Some diseases are particularly to be feared, such as gastro - intestinal disorders, respiratory or skin diseases, whose responsible agents can also be responsible for TIAC (Collective Food Toxi- Infections) and develop on and in food. All these sick people must be treated and, of course, kept away from food handling areas.

The problem is more important when it comes to healthy carriers. And they are frequent: intestinal carriers of salmonella or shigella, cutaneous carriers of staphylococcus

Regulations require that no person known to have a disease that may be transmitted through food is permitted to work in a food handling area in any capacity where there is a risk of direct or indirect contamination of food by pathogenic organisms.

Any member of staff called upon to handle foodstuffs must have been declared fit to do so. The person in charge of the establishment ensures that this aptitude is medically certified, each year, in accordance with the specific regulations in force. It is therefore specified that any person who has to handle foodstuffs must present a medical certificate declaring him/her fit for such work (employment certificate).

This certificate must be renewed every year. In addition, such a certificate must be provided after an absence of more than six months.

Microbiological screening examinations should be performed in the following three cases:

- If the clinical examination suggests a dangerous condition.

- If the microbiological analysis of the foodstuffs suggests contamination by the staff.

- When returning to work after sick leave for digestive tract or respiratory disease.

The occupational physician will have to record the results of his examinations on a numbered health register, entrusted to the operator.

Skin conditions are particularly dangerous when they involve the hands, arms, neck or face. These can be poorly treated wounds or others. Colds and sinusitis should be associated with these conditions. If a worker has such a condition, he or she can only work in a job that is compatible with the condition, such as packing, shipping, cleaning non-food areas, etc.

The plant must have an infirmary to provide first aid.

2. Body cleanliness:

All persons working in a food handling area must maintain a high standard of personal cleanliness. This means that staff must have the opportunity to shower regularly and frequently.

Most importantly, workers must wash their hands so that they are always clean. The person in charge of the establishment will have to verify this bodily cleanliness.

Washing is the basic hygienic operation. It must become a reflex in a number of situations:

- Upon return to work

- Leaving the bathroom: Hand washing after leaving the bathroom is crucial.
  In fact, there can be 10 billion germs per gram of stool, which can be transferred to food through hands and clothing.

- After blowing your nose

- After handling contaminated raw materials (earthy vegetables, eggs, food packaging, waste, etc.).

- Before handling prepared and contaminated food (meats, sauces, etc.),

- Between each separate operation. The change of activity is a factor favouring cross-contamination. Therefore, whenever possible, we should try to separate "clean" and "dirty" activities.

- Of course, before washing your hands, you should remove all rings and bracelets. The nails should be cut short, always clean and never varnished.

Although it may seem obvious, it is important to remember that handwashing stations should not be located inside a risk area. They should be located indoors in a specific room, most often in an airlock, with hand washing as part of the protocol for entering the area.

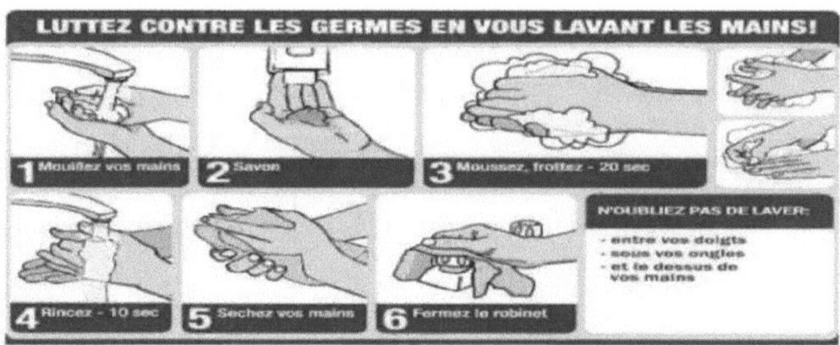

Etapes de lavage des mains

It is preferable that the sink be designed to allow for thorough cleaning:

- In non-porous material

- With an adapted size and design

- No stagnation zone

- No areas inaccessible to cleaning

In simple cases, the sink faucet can be manual, but if the risk of contamination is greater, the water supply must be ensured by a pedal, a lever at the elbow or an automatic system such as a photocell.

The bacteriological quality of the water distributed to most sites is undeniably good. However, this quality can be altered, in significant proportions, depending on the configuration and maintenance of the network specific to the establishment. It seems necessary to carry out a regular microbiological control of the water (including the first milliliters emitted).

The main difference between soap valves is the control system.

The collective hand towel should no longer exist, it must be replaced by one of the following three processes:

- Pulsed hot air drying: This process has the major disadvantage of diffusing skin flakes and associated microorganisms into the environment and resuspending the particles deposited on the floor and walls,

- Disposable paper towels: This process is based on the distribution of paper or non-woven towels in folded or rolled format. It has the disadvantage of not meeting all the desired characteristics (resistance, linting, absorption capacity), unless it is of excellent quality, but in the latter case, quality is synonymous with high cost.

- The textile towel - single use in automatic reel: This textile must have undergone at the end of washing an effective bacteriostatic and fungistatic treatment. We will quote for memory, the processes which will not have to be met any more including in the field "general public", of the traditional cloth, always soiled and wet and seat of a very important proliferation of microorganisms.

The fabric, whatever its nature, is an excellent support for bacteria. When it is moist and impregnated with nutrients, it constitutes a real culture medium where microorganisms develop abundantly.

3. Cleanliness of clothing:

Clothing:

In general, clothes are all the more contaminated as they have been worn for a long time and have been in contact with a polluted environment. Their flora reflects the flora of the individual and that of his environment. The work clothes will be chosen in white fabric to show the state of cleanliness. They will be changed every day.

The oral-nasal mask or the bib:

The oropharyngeal cavity can be an important source of contamination, which is why wearing a single-use bib is mandatory. Note that it is only effective if it covers the nostrils.

The headdress:

The hair can intervene indirectly or directly. The direct intervention is carried out by the hair which falls in the product or in a packaging. It will be found by the consumer in the food he buys. The indirect intervention concerns the microorganisms that they support. The spores of *bacillus* and *micrococcis* are among the most frequently encountered.

The only protective method is to wear a cap. In order to be effective, the cap must envelop the entire head of hair. In the case of very long hair, it is essential to pinch the hair with two clips to prevent it from moving.

Wearing a beard is not recommended for people who are in constant contact with unpackaged foodstuffs. No matter how well cared for, a beard still carries a lot of bacteria.

Wearing gloves:

If the personnel wears gloves, it will be necessary to disinfect them by regularly passing them through an antiseptic solution placed in a tank near the work station.

The antiseptic will usually be a chlorinated alkaline or a solution containing a mixture of alcohols.

It should be pointed out that wearing gloves is often an illusory protection because the flora that develops on the enclosed, moist skin at the right temperature escapes through the inevitable perforations

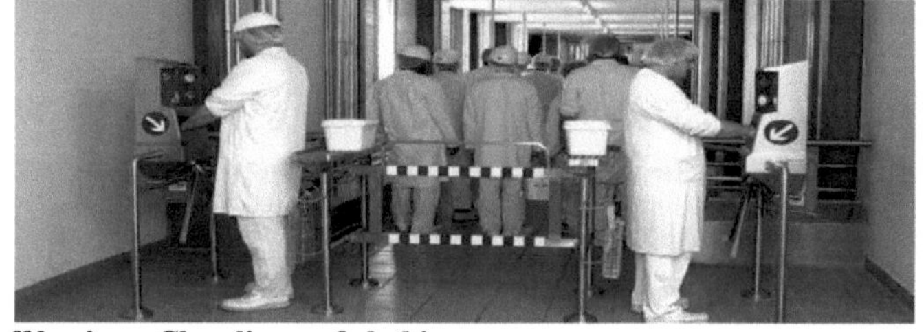

Staff hygiene: Cleanliness of clothing

Shoes and foot baths:

A distinction must be made between street shoes and work shoes. The former should never be brought into the production area.

It is therefore necessary to remove the shoes, upon arrival at the factory, in

suitably located changing rooms which do not oblige the personnel to cross the workshops with street shoes. Therefore, the personnel thus shod must not leave the factory without putting on their street shoes.

The boots must be the object of the greatest attention. At the end of the work, before removing them, they will be brushed under a jet of water.

The installation of footbaths at the entrances of the workshops is recommended, with an appropriate antiseptic solution.

4. Personal behavior:

Food handlers should avoid behaviors that could result in food contamination:

ɟ Smoking: Cigarette ash that falls spontaneously can "dissolve" in liquid food or soil solid food. Cigarette butts that smokers don't know what to do with and that they throw on the ground are a factor of dirtiness and soil contamination.

ɟ Spit

ɟ Chew or eat

Sneezing or coughing near unprotected food: studies have shown that a sneeze results in the emission of 20,000 to 40,000 droplets ranging in size from a few micrometers to 1 mm, with a critical splash distance of up to 1 m.

Visitors admitted to the manufacturing, processing or handling areas will be subject to the same instructions as the personnel. They must be provided with a pair of boots, otherwise they will have to be satisfied with observing the workshops from a side gallery provided for this purpose.

5. Training:

Training is of fundamental importance in any food hygiene system. All persons whose activities relate to food must be trained and/or instructed in hygiene and supervised, otherwise they pose a potential threat to the safety and suitability of food

Food business operators who come directly or indirectly into contact with food should receive training and/or instruction in food hygiene at a level appropriate to the operations they perform.

Awareness and responsibility:

All personnel should be aware of their role and responsibilities in protecting food from contamination and spoilage. Food handlers should have the knowledge and skills to handle food hygienically. Those handling strong cleaning agents or other hazardous chemicals should know how to handle them safely.

Training programs:

Factors to consider in assessing the level of training required include:
- The nature of the food, particularly its ability to support the growth of pathogenic or decaying microorganisms.
- How food is handled and packaged, including the risk of contamination.
- The extent or nature of further processing or preparation before final consumption.
- The conditions under which the product will be stored.
- The expected time before consumption.

Chapter 3: Operational Hygiene

1. Reception of raw materials:

The reception of raw materials is a key position in the food industry.

- Check that the product corresponds to the "specifications" on the documents and by controls
- Reject non-conforming products or damaged packaging. The major contaminants of raw materials are (1) rotten and moldy, (2) soil (plants), (3) feces (animals).
- Do not add contamination
- Separate the different deliveries
- Store immediately under the correct conditions

     Raw materials are a key source of contamination. Hazards can be:

- **Microbial**: foodstuffs initially contaminated by pathogenic germs (*Salmonella, Clostridium perfringens, Staphylococcus aureus, Listeria monocytogenes...*), their toxins or metabolites. In addition, in some cases, constituents of raw materials can be degraded by spoilage bacteria, resulting in spoilage residues that are toxic to the consumer.
- **Chemical**: foodstuffs containing chemical residues introduced upstream during their preparation at the supplier's for example.
- **Parasites**.

- **Mechanical**: residual foreign bodies that can be ingested by the consumer.

Preventive measures are based on:

- A good knowledge of the purchased products, by elaborating product specification sheets, including all the characteristics necessary to identify them and evaluate their quality.

- Each characteristic will, if necessary, be accompanied by the measurement method and the accepted tolerance range. Of course, the elaboration of these specification sheets will have to concern in priority the products for which the risks* are the highest.

- The product specification sheets are based on existing standards (CODEX ALIMENTARIUS, AFNOR).

- The conformity of the purchased products will be verified upon receipt, systematically for certain criteria, for others on the basis of sampling or from documentary evidence requested from suppliers (results of analyses, audit reports, various records...).

- A good knowledge of the suppliers selected and referenced after verification of their approval by the Departmental Veterinary Services (when required for the category of products considered), the existence of an effective food safety and quality assurance policy and possibly the certification of their company according to ISO 9000 standards will be points to be privileged when choosing suppliers.

2. Handling of raw materials:

During transport, foodstuffs can be damaged and contaminated by germs from :

- Other commodities transported

- Deconditioning

- The transport vehicle itself (interior lining, lashing equipment, pallets, hooks, etc.)

- Transportation personnel.

Upon receipt, new sources of contamination may alter the product.

Thus the contamination can come from :

- Handling personnel,

- Reception equipment (handling, weighing...),

- Reception area,

- Of the environment if the foodstuffs have to transit through the exterior.

The bacterial load of the products can increase during transport and the freshness of the product can be modified by :

- Non respect of the temperature of the foodstuffs,

- Alteration of the packaging or inappropriate packaging of foodstuffs,

- Delivery of products with an expiration date that is too short or even exceeded.

The microbiological quality as well as the freshness characteristics of the foodstuffs can also be altered if the reception modalities lead to a rise in the temperature of the products which is incompatible with the required optimal conservation conditions.

Preventive measures include:

- Control the conditions of transport: cleanliness of trucks, integrity of packaging, temperature of the enclosure. The frequency of these controls will take into account the nature of the products and the previous results of the supplier.

- Control the delivered products: integrity of the packaging, conformity of the labelling, in particular the remaining shelf life of the product before the BBD or the UBD, product temperature.

- Given the uncertainty inherent in any measurement, a difference of 3°C will not systematically lead to the rejection of the goods delivered. On the other hand, an action of sensitization and information will be carried out with the carrier and the supplier.

- To ensure a fast reception of the products: this implies an organization of the schedules of delivery in order to avoid the congestion, the respect of a logic in the reception, the control and the storage of the products:

- Priority to foodstuffs,

- Priority to foodstuffs to be stored at the lowest temperatures,

- Wet" product before "dry" product,

- Unpackaged products before packaged products.

- Identify and isolate non-conforming products that are not immediately returned by the supplier.
- Keep records of these checks.

These preventive measures require resources:

- Training of the reception staff, who must have in particular the information on the temperatures in reception by category of products and the necessary instructions to carry out the controls and to act in the event of noted anomalies.
- Provision of reliable and regularly checked probe thermometers.
- The storage of foodstuffs can be the cause of their contamination. We can fear in particular:

In a refrigerated or negative cold chamber

ɟ The coexistence of bare, packaged or simply wrapped foodstuffs

(transmission of telluric germs such as *Listeria* from cardboard boxes).

ɟ Storage of products directly on the ground.

ɟ Poor storage of products, leading to settling and deterioration of packaging (particularly vacuum-packed products)

ɟ The simultaneous presence of packaged (cartons) and wrapped (plastic film of vacuum-packed meat) foodstuffs: the packaging may be contaminated and the contamination will occur later during unpacking.

ɟ The simultaneous presence of foodstuffs of incompatible sanitary levels (fruits, raw vegetables and unpackaged meats for example).

J The presence of foodstuffs that have been "returned" (semi-finished foodstuff, intermediate product in the course of manufacture). In this case, the foodstuff and its container can be the vector of cross-contamination between the storage area and the preparation areas.

J The presence of foodstuffs that do not belong to the establishment (personal foodstuffs, pharmaceutical products, etc.). This type of food should be limited as much as possible (if this type of food is introduced, it should be kept separate from the other products).

<u>In dry food reserve</u>

Contamination of products by pests (insect larvae introduced by a foodstuff in the dry storage) and rodents is added to the previous sources of contamination*.

The ideal method is AMER for germs.

A- Contribution- To avoid the contribution of microbes, think of the **"5 M"**. The main factor= manpower.

An automated operation is less risky than a manipulation.

Mechanical operations make the entire product accessible to a contaminant (*e.g. slicing, chopping, grinding, mixing*). Therefore, great care must be taken to keep the machines clean.

M- Multiplication- Bacteria can only multiply if they have time, and the risk decreases with time and temperature: refrigerated and organized workshop, cold chain. The composition of the product can limit the microbial multiplication (water (aw, salt, sugar), pH, nitrate...).

E- Elimination- Heat treatment (cook at 70oC, pasteurize, or sterilize the food) and Cleaning / disinfection (equipment, premises, personnel).

R- Recontamination- As soon as possible, we pack (a packaged food is protected).

Chapter 4: Cleaning and disinfection

1. Definitions

The cleaning and disinfection operations allow to preserve both the health and the well-being of people, to maintain the premises in a sanitary state. For the particular case of the food industry, the quality of cleaning strongly conditions the quality of the work and the results obtained.

To clean: it is to eliminate the soiling = to make the surface clean.

Disinfecting: is to temporarily reduce the number of germs by destroying the pathogens (unlike sterilizing which permanently eliminates the germs). It is not possible to disinfect a dirty surface, so it is better to clean without disinfecting than the opposite.

Soiling is synonymous with **Dirt**: It is an undesirable contribution on the surface and or inside the support and which alters certain characteristics of aspect or touch of the clean surfaces. The stains can be of very diverse nature, according to: their dimension (of some micro meter to several mm), their density, their consistency, their chemical nature.

2. Different factors that intervene in the cleaning and disinfection operations:

Four variable factors are to be taken into consideration: Time, Temperature, Mechanical action and Chemical action = TACT. A cleaning operation is represented by means of the factorial circle, known as the "SINNER circle". The decrease in importance of one factor must be compensated by the increase of one or more others if the final quality of the cleaning is to be maintained.

- ◆ **Time:** Time is often lacking. However, it is obvious that most chemical reactions require a certain amount of time to be completed in a satisfactory manner.

- ◆ **Temperature:** Hot water is rarely available on the cleaning site. In addition, many materials can hardly or not at all withstand high temperatures. However it is known that the removal of greasy dirt is facilitated by heat.

❖**Mechanical action:** It is implemented by many cleaning machines but must be limited in some cases because of the risk of deterioration of the material constituting the surface to be cleaned; in any case the mechanical action is insufficient on its own.

❖**Chemical action:** It is on it that most often rests the essential of the operations of cleanings. the more effective the chemical action will be the more it will be possible to reduce the 3 other factors. The design of a cleaning product must take into account in addition to the results sought, the interactions due to coatings and machines.

The products must be able to dissolve all kinds of dirt without harming people and their environment and without damaging the surfaces. There is currently a wide variety of products whose effectiveness is generally excellent, provided that they are used properly.

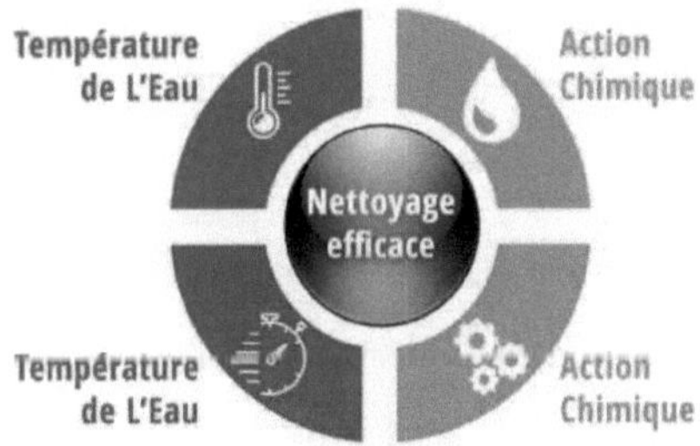

SINNER Circle

N&D will be done more "thoroughly" in the ultra-sensitive areas, "at high risk" for the food: bare food, material is in contact with it, food is fragile (high aw, neutral pH), or is going to be stored for a long time, and without subsequent treatment. A complete N&D protocol should be applied, in 7 steps. At lower risk levels (e.g. storage of packaged food) a lighter protocol is applied (cleaning only without disinfection).

3. N&D protocol in 7 steps:

1/Tidy, 2/Prewash, 3/Clean, 4/Rinse, 5/Disinfect, 6/Rinse, 7/Dry

Sometimes simplified to 5 steps using a specific disinfectant detergent (for N&D at once)

1- Tidy-Protect: dismantle, unplug, take out food &waste....

2- Prewash with cold (milk proteins) or hot (fats) water and scrape

3- Clean. The goal is to loosen and suspend the dirt (Apply the four efficiency factors previously mentioned TACT)

4- Rinse to remove soil and detergent (which inhibit disinfection) with hot water

5- Disinfect in general cold and in 10 min.

6- Rinse with cold drinking water to remove disinfectant residue

7- Dry and store to limit recontamination   (Listeria&

Pseudomonas). It is left to drain and dry spontaneously.

4. Cleaning agents:

4.1. The different cleaning agents:

The essential constituent of the class of cleaning products is always a surface-active agent, associated with other components whose nature depends on the particular destination of the product considered. A detergent contains in addition to the surface-active agent, complexing agents, buffer mixtures and various additives.

Surfactants:

To understand the role of surfactants, we must first know what is the surface tension of water. It is easy to observe that a drop of water placed on a smooth and impermeable surface (e.g. glass slide) spontaneously takes an almost perfect spherical shape. This spherical shape is the one that allows to obtain the smallest possible surface compared to the volume of the water drop. It is the tension that is responsible for this phenomenon.

The disadvantage of surface tension is that it prevents water from coming into close contact with the surface to be cleaned and the dirt to be removed. It is sometimes said that water does not wet. To overcome this disadvantage, it is necessary to modify the surface state of the water by adding a product called surfactant. Among other effects, surfactants reduce the surface tension.

<u>Mode of action of surfactants</u>:

Surfactants are organic molecules that have a hydrophilic polar part and a hydrophobic apolar part (hydrocarbon chain). Whatever the type of surfactant used, the process of action against greasy dirt remains the same: the lipophilic radical flees the "water" element in which it is not soluble and will fix itself on the dirt and greasy dirt.

- Wetting and spreading power: fixing the apolar group on greasy soils.

- Penetrating power: partial disintegration of dirt

- Penetrating and dispersing power: final disintegration of the dirt into fragments and suspension in water.
- Emulsifying and anti-redeposition power: obtaining fatty particles enveloped by surfactants. They are micelles, the apolar greasy dirt becomes polar on its periphery and can then be washed away with water.

<u>The classification of surfactants</u>:

Surfactants are classified according to the electric charge carried by the organic radical in aqueous solution. We distinguish thus:
- **The anionic surfactants** (they carry a negative charge): the hydrophilic part is generally an "acid" radical, as in the case of soaps, they ionize in water, releasing an ion that has a surface-active power.
- **Cationic surfactants** (they carry a positive charge) are mainly quaternary ammonium salts.
- **Non-ionic surfactants** (without electrical charge)

- **Amphoteric or ampholytic surfactants** (the charge is either positive or negative depending on the pH and the medium of use)

<u>Acidic and basic components:</u>

They are present in many cleaning products.

<u>Acidic components:</u>

We distinguish between mineral acids and organic acids. Among the mineral acids authorized for use in foodstuffs are

- **Hydrochloric acid:** only used cold, not often used because of the risk of corrosion even on stainless steel, risk of handling
- **Sulfamic acid:** weaker than the previous one, its use is less dangerous for the operators.
- **Phosphoric acid:** it is used in the composition of most products "acidic detergents" used as a descaler

- **Organic acids:** They are interesting because they are neither dangerous nor corrosive. Example: Acetic acid: this weak acid is commonly used in dilution (2 to 8%) in rinsing water to neutralize basic product residues.

<u>Basic components:</u>

- **Caustic soda** (sodium hydroxide) is an extremely corrosive and dangerous product, to be used with care in a pickling operation. It destroys many organic stains by saponification and facilitates their solubilization.
- **Caustic potash**: It is less used than soda for two main reasons : its higher price and its higher molecular weight (more product for an identical basicity).
- **Ammonia:** Helps to remove greasy dirt. It is less used in its free state because of its volatility and the risks and odors that it generates.
- **Silicates: they are** good surfactants which, thanks to their basic pH (10 to 12), hydrolyze organic dirt. They are less aggressive than the previous ones.
- **Phosphates**: they act mainly as a sequestering agent.

<u>**Sequestrants (or complexing agents):**</u>

The water distributed by urban networks is hard, even very hard. However, the wetting power and the emulsifying power of most surfactants are strongly reduced by the presence of calcium and magnesium ions in the dissolved water (washing solution). This leads to poor results in cleaning operations, which forces to overdose the products to compensate for their loss of efficiency. It is therefore necessary to sequester or complex these calcium and magnesium ions with molecules: this phenomenon is called "chelation" or "sequestration".

Polyphosphates and metaphosphates can fulfill this role. In addition, these molecules participate in the anti-redistribution function of dirt. They also have good fat emulsifying properties. Unfortunately, polyphosphates have the disadvantage of being highly polluting. Laundry manufacturers have developed substitutes: orthophosphates and sodium silicate.

The most common **complexing agents** currently found in elaborate products for cleaning operations are of two types:

- Sodium tripolyphosphate

- EDTA (ethylene diaminetetraacetic acid disodium salt)

3.1. Choice of cleaning products:

A number of factors can directly influence the effectiveness of cleaning products. That is why it is necessary to consider the following before choosing a product

- ◆ The nature of the stains (emulsifiable, swellable, soluble.)

- ◆ The nature of the surface on which the stain is deposited (structure, energetic property, thermal stability...)

- ◆ The water used (scaling, aggressive, polluted)

- ◆ The nature of cleaning

◆ Biodegradability

A detergent is only suitable for industrial use if it is cheap and relatively easy to use. It should also be noted that when selecting cleaning and disinfecting products for simultaneous use, compatibility must also be taken into account.

3.2. Properties of detergents:

◆ **Wetting power :** The surfactant molecules (detergent) are inserted between the water molecules and break the surface tension forces. This increases the solvent and rehydrating power of water.

◆ **Emulsifying power :** Grease particles and other fatty soils are attracted to the apolar parts of the detergent molecules while water is attached to the polar parts. The dirt particles are thus developed from detergent molecules, which are themselves bound to the water. The removal of dirt from the support becomes easy and the fatty particles float in the liquid without falling back: it is an emulsion.

◆ **Dispersing power:** Insoluble particles must be kept in suspension in the water to prevent them from redepositing, which would greatly reduce the final cleaning quality.

◆ **Foaming power:** Foam can be considered as an emulsion of air in liquid. Strongly agitated surfactant solutions stirred or compressed produce foam in variable quantities. The physical balance of the forces that tension the surfactant bubbles is very subtle and the parameters that vary the foaming power of a detergent must be controlled with maximum precision. Indeed, in some cases, foam is useful because it allows better contact between the detergent and the surface to be cleaned. In other cases, the foam is particularly annoying in the cleaning operation. This is why the detergent manufacturer is committed to controlling the foaming power with the utmost precision.

◆ **Rinseability:** This is the ease of elimination by rinsing with clear water of the product loaded with impurities. It is also a very important criterion of choice of the detergent product in many cleaning operations.

◆ **Biodegradability:** It is the property that presents a compound to be able to undergo a chemical degradation under the action of living organisms present in the natural environments and which use these compounds as food. The legislation in force requires detergents to be more than 90% biodegradable.

A good detergent should be :

- Effective at recommended dosages

- Non-toxic to the user at the usual dilutions of use

- Non toxic for the environment (biodegradability>90%)

- Non-aggressive to cleaned surfaces and cleaning machines

- Easy to dilute

- Easy to rinse

- Not very sensitive to water hardness

- Packaged in easy-to-handle security packaging

- Adjuvants: Clearly, legibly and completely labeled

5. The different disinfecting agents:

Many disinfectants are more or less active than others or better tolerated. Their choice is sometimes difficult, often because of a lack of information on their activity. Sometimes the active ingredient can be used pure or diluted, usually in water. The choice of a particular disinfectant depends on the environment to be disinfected, the availability of the product, its cost, etc.

Disinfection works in three steps:

- Attachment to the cell wall

- Blocking of cytoplasmic membrane enzymes

- Involvement of intracellular   constituents

Among the existing products are:

5.1. The main disinfection agents

Chlorine and its derivatives

- hypochlorite (bleachNaOCl),

- chlorine peroxide (ClO_2),

- hypochlorous acid (HClO).

Microorganisms are destroyed either by oxidation of certain metabolites or constituents of these, or by formation at the expense of the latter of chlorinated substances toxic to the cell, or by inhibition of internal enzymatic processes

Aldehydes

- Miscible with water, alcohol and other organic solvents

- Broad bactericidal and fungicidal activity

- Toxicity depends on the structure of each product

Amphoteric compounds

These compounds contain an anionic and a cationic group. They have the following properties:

- Soluble in water and biological media

- Broad spectrum of activity for a pH around 7

- Low toxicity

- Have a detergent power.

Other compounds

- iodine: not very soluble in cold water, strong odor, corrosive to some metals, very limited use in the food sector
- Ozone: the most powerful oxidizer, broad spectrum of action, burns pollutants such as detergents, fermentation residues, phenols and all organic debris, leaves no trace, no odor, no toxic residue.
- Hydrogen peroxide or hydrogen peroxide: same advantage as ozone. It is used to sterilize flexible packaging materials for the aseptic packaging of UHT milk.
- Peracetic acid: this product is a combination of hydrogen peroxide and acetic acid. It is a very active agent, on bacterial spores or bacteriophages, effective at low temperature, corrosive on materials and harmless to the environment. It rinses easily and does not even require rinsing. It is one of the most effective disinfectants currently available. It is widely used in breweries and dairies. In combination with alcohols, it is used for the sterilization of packaging in aseptic packaging machines.

5.2. Properties of disinfectants:

To ensure proper disinfection, the product should meet the following requirements:
- To have a very broad spectrum of effectiveness and to be able to destroy all microbial species with the same speed
- Be usable at low concentration
- Be safe for users even at high concentrations
- Be chemically stable
- Be without corrosive action on materials

-Be easy to rinse

-Have an equal efficiency in the presence of interfering substances, i.e. in the presence of organic matter

No disinfectant has all these qualities, and in practice, more modest criteria are defined according to the use to be made of it.

6. Industrial realization of the operations of cleaning and disinfection:

6.1. Manual industrial realization :

<u>Brushing</u>:

Many manual cleanings are done with brushes. This brushing operation is sometimes necessary for the dismantled materials of certain industrial equipment (parts of packaging machines, cutting machines, pushers, filter plates, etc.). Manual brushing tends to be replaced more and more in new installations by automated techniques.

<u>The immersion or soaking process</u>:

The process by immersion or soaking consists in leaving the material in detergent and disinfectant solutions. This process was widely used for molds, cheese racks, small salting equipment, filter press cloths in the beverage industry, etc. However, this technique is tending to be replaced by continuous washing tunnels, which make it possible to automate production lines and considerably increase production rates. On the other hand, it is also possible to automate the loading and unloading of the material to be treated in the soaking baths, but this process is less used than the washing tunnels. On the other hand, the soaking in disinfectant solutions of dismountable materials is still widely used, this operation follows the cleaning operation carried out in this case most often by manual brushing.

6.2. Mechanized industrial realization

<u>Foam cleaning and disinfection - Pressure rinsing</u>:

The need to improve the hygienic quality of food products, the continuous growth of labor costs, the aspiration to better working conditions has led to the mechanization of cleaning. Apart from closed circuits, pipes, tanks, etc., which are washed by means of a central cleaning system, the most commonly used cleaning technique today consists of the application of a detergent in foam form followed by rinsing with pressurized water. Initially used in the meat industry, foam cleaning with pressurized water rinsing is rapidly spreading to the entire food industry.

Cleaning in place:

Cleaning In Place (CIP) is now a widely used technique in the food industry (a good example is the dairy industry), for the cleaning of closed systems composed of networks of tubular connections linking different equipment and tanks through the circulation of water, detergents and/or disinfectants. All these operations do not require any dismantling.

A cleaning cycle generally consists of the following sequences:

- A push with water and / or air in order to reduce the loss of product, facilitate the cleaning and reduce the importance of effluents.
- An initial rinse with water as soon as possible to avoid drying out of residual dirt. The rinsing water leaving the installations must be as clear as possible
- Cleaning by circulation of a hot detergent in a closed loop or without recovery in a tank
- Intermediate rinses with or without recycling. The objective is to eliminate

any trace of the detergent which can be measured either by pH-metry or conductimetry,

- A possible passage of a second detergent. Several types of detergents are sometimes used consecutively with intermediate rinses (alkaline, rinse, acid, rinse)
- Disinfection. In many cases, the disinfectant is injected into the circulating flow of rinsing water followed by a stop time where the system will remain in contact with the disinfectant
- Final rinse with drinking water.

- The time/temperature parameters of each cleaning sequence are specific to the equipment or the type of installation to be cleaned and generally to the type of transformation process.

Cleaning unit in place

7. Control of the effectiveness of the cleaning and disinfection operations:

It is possible to distinguish two main categories of methods, namely direct and estimated methods.

7.1. Direct methods:

They will consist of the total recovery of the residual microflora present.

Rinsing: This method is mainly used for circuit controls. Disinfection controls of pipes, tanks, bottles, injection pumps are all indicated by this method. It is then a question of washing a given circuit with a sterile nutritive rinsing liquid. The entire liquid or an aliquot of the total known volume is processed for inoculation. The disadvantage of this method is that it does not allow a precise localization of a contamination resistant to disinfection (biofilm in a pipe for example). However, it gives very good results on the localization of circuit sectors.

Casting: Casting is specifically used for the control of small containers (boxes, tubes, bottles). It consists in pouring an agar in a homogeneous way maintained in supercooling inside the surface (rolling tube). Then the container is incubated. Although very simple, this technique requires dexterity on the part of the operator which can only be acquired through long practice.

Both methods will provide both quantitative and qualitative results, which will be expressed as the number of germs per cm^2 of surface examined.

Bioluminescence or ATP metry: This is the most recent method currently available to the microbiologist. It is located halfway between direct methods and estimated methods. It is based on the principle of bioluminescence whose biochemical mechanisms involve an enzyme (oxidase) and a substrate (ATP), a universal intermediate of the living cell. To simplify: a molecule of ATP in a luciferin / luciferase complex will give rise to a photon. The quantity of light emitted will be proportional to the quantity of ATP present and consequently to the quantity of living cells. However it will be necessary to take into account the differences in concentration of ATP present in the various living cells (a mold contains nine times more ATP than a bacterial cell).

It is thus enough to proceed to a known washing of a surface whose dimension is exactly determined, then to ensure an exhaustive recovery of the rinsing liquid and finally to subject this one to reaction of bioluminescence. A quantitative response (number of bacterial cells per cm^2) will be obtained. There is still a problem to be dealt with, which is that of freeing the rinsing liquid from the non-microbial ATP that it may contain (organic contaminants). This method is fast and elegant but can only inform the microbiologist about the total residual flora.

7.2 Estimated methods:

These are most often methods acting by direct contact between an agar surface and the surface to be tested.

Swabbing: After having determined a field of experimentation of exactly known surface (generally 10 cm^2). This is done with circular or rectangular tools that are easily sterilized (stainless steel). A complete swabbing is then carried out inside the surface thus delimited. Then the swab is either:

>**Inoculated** by depletion on a suitable agar medium

>**Dissolve**　(alginate buffer) in an exactly known volume of nutrient liquid

>**An** aliquot part is not very precise but keeps the quantitative and qualitative
aspects

Chapter 5: ISO 22 000 vs 2005

1. Definition of the ISO 22 000 standard:

The ISO 22000 standard specifies the requirements of a food safety management system, which is a coherent set of activities designed to enable the management of an organization to ensure the efficient and effective implementation of its policy and improvement objectives.

All companies, regardless of their size, sector of activity or local presence, have more or less formalized their management practices over time.

The existence and control of a food safety management system can help the company give stakeholders confidence that there is a management commitment to implement its policy in its decision-making processes and the information and measurement system to judge it. The potential benefits of implementing an effective food safety management system include

- the assurance provided to the various actors of the food chain of a more effective and dynamic control of the dangers related to the safety of foodstuffs,
- the ability to consistently deliver safe finished products that meet both agreed-upon customer requirements and food safety regulatory requirements,
- assurance to stakeholders of transparency in its organized and targeted communication between partners,
- the implementation of a structured approach that involves all personnel in a continuous improvement process.

It concerns all the actors of the food chain: the organizations directly involved (producers / processors / distributors), or indirectly involved in the chain (suppliers of materials

of packaging/cleaning products...).

The ISO 22000 standard is based on the principle of the Deming wheel and its continuous improvement loop of the PDCA type (Plan, Do, Check, Act) which is now recognized as a simple and universal managerial principle. Its application to corporate management systems has been widely applied in recent years after having proven its effectiveness in Japan.

The structure of the ISO 22000 standard takes into account the provisions contained in the ISO 9001 standard in order to allow a perfect compatibility and complementarity with the various management reference systems currently used by companies. It is based on four main blocks closely linked :

- Management's responsibility,

- Resource management,

- Planning and realization of safe products,

- Validation, verification, and improvement of the system.

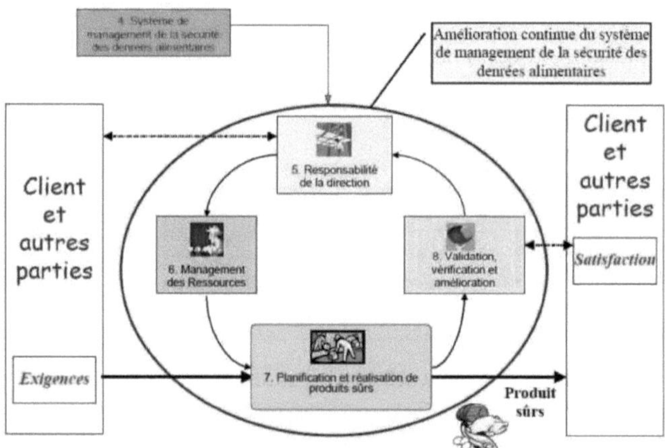

Structure of the ISO 22000 standard

The figure above illustrates the model of the approach taken in ISO 22000 around the four main blocks. The food safety requirement is integrated into a structured, effectively managed management system that fits seamlessly into the overall management activities of an organization.

2. Key elements of the ISO 22000 standard

The ISO 22000 standard is applicable to all actors of the food chain. Beyond the quality requirements with which the ISO 22000 standard is perfectly convergent, this standard specifies requirements including 5 elements that are recognized as essential to ensure food safety at all levels of the food chain: the systemic approach, interactive communication, traceability, prerequisite programs (PRP) and HACCP principles. These elements are integral to the requirements of the standard (see figure below).

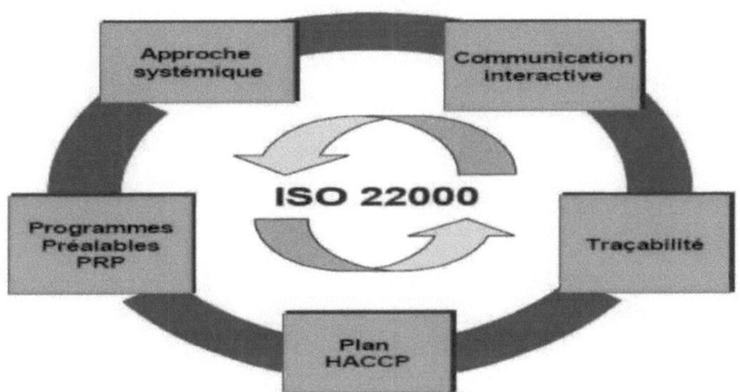

Key elements of the ISO 22 000 standard

Chapter 5 of ISO 22000 deals with management responsibility. The commitment of the management must not be limited to a single written or oral act but must be translated by a strong and concrete implication on the ground. The commitment of the management and its involvement is an important criterion for the improvement of the company's performance.

This chapter presents the requirements for management in a dynamic cycle from food safety policy to communication and contingency response in emergency situations. Emphasis has been placed on the notion of the company being part of a chain: the paragraphs dealing with the communication of food safety data, both upstream and downstream, have been drafted with a constant concern for the balance between useful transparency and confidentiality of information from each entity in the chain.

Interactive communication between the different actors (see figure below) at all levels of the chain is essential to ensure that all relevant hazards are identified and properly controlled.

The different actors of a food chain

Chapter 6 focuses on the provision of resources necessary for the implementation of the food safety management system and its maintenance. The emphasis is on the human resources component, thus affirming the essential role of the

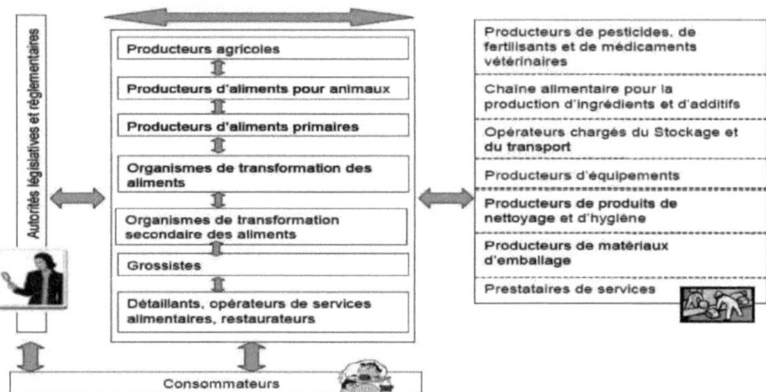

men and women of the company.

Chapter 7 deals with the planning and implementation of safe products. It dynamically links the prerequisite programs with the implementation phases of a HACCP plan as described by the *Codex Alimentarius* Commission. The PRPs are the key points of good hygienic practice that are considered systematically in order to deduce what falls under actually a CCP (Critical Control Point) identified in the design and modification of the HACCP plan.

In line with the provisions of Article 18 of the new European Regulation 178-2002 laying down procedures in matters of food safety, this chapter also requires the company to establish a traceability system.

The fourth block of ISO 22000 (Chapter 8) is the feedback loop of the food safety management system. It is to ensure that the results are consistent with the objectives set for food safety. The processes necessary to validate, verify and improve the system must be implemented.

The implementation of the ISO 22000 standard will allow a company to integrate the requirements of its customers and the regulations in terms of food safety in a global approach where the articulation between PRP and HACCP programs is done in a dynamic way and with a concern for continuous improvement and transparency.

3. ISO22000 certification

Certification to ISO 22000 can be fully compatible with management system certification. Indeed, the conception of the ISO 22000 is coherent with the ISO 9001, itself compatible with other international standards (of which ISO 14001/ certification of the management systems environmental).

Certification by an independent and competent third party is a legitimate means of gaining international recognition for the food safety management approach.

Why invest in a certification process?

- The ISO 22000 certification or assessment is an external, impartial and rigorous look at the company's organization. It allows the identification of the progress areas necessary to improve the food safety management system.

  - Certification / evaluation procedures are an indisputable proof of its concern for transparency and a guarantee of confidence towards its customers. They constitute additional assets for companies in the face of competition.

  - Finally, ISO 22000 certification, because of its status as an international standard, will be a passport to export, allowing you to have your food safety management system recognized.

ISO 22000 auditor qualification and certification process

The sectoral selection of ISO 22000 food auditors is reinforced by a selection according to :

- knowledge of the auditing profession,

- knowledge of the ISO 22000 standard and the audit methodology validated by an exam,

- practical knowledge of the product categories to be audited,

Throughout the certification process, emphasis is placed on the evaluation of practices in the field. The preparation of the audit is done by an auditor adapted to the size of the company and competent in the field of activity of the company. The preparation on site aims to :

- Visit the site in depth and understand the food safety issues within the company,

- Preparatory review of documents (HACCP plan, preliminary program and flow charts)

- Develop an audit plan adapted to the organization of the company and its possible constraints (activity schedules, availability of people to meet...).

- Realization of the audit on site of the company to be audited

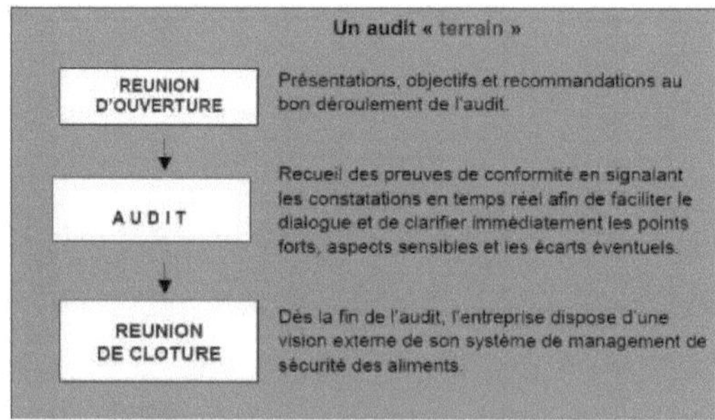

Phases of the certification audit

4. <u>HACCP method</u>

Definition:

HACCP = Hazard Analysis - Critical

Control Point

The HACCP is a method to identify all the hazards associated with a food, then control them during manufacture by systematic and verified means.

An accumulation of technical means cannot guarantee safety. A rigorous approach is needed to adapt the means to the defined objectives (safety). The HACCP proposes a structured, responsible, specific, preventive, creative method, which integrates the already known means: to define the necessary and sufficient self-checks.

Compared to quality assurance (aims at all the components of quality), the objective of HACCP is to ensure the safety (harmlessness) of food: it is a safety assurance plan.

The HACCP method was invented by Nasa to avoid the TIAC of astronauts.

Principles of the HACCP method

- **biological danger**: parasites, viruses, pathogenic bacteria, microbial alteration, toxins
- **chemical hazards**: allergens, pollutants, drug or pesticide residues, contaminants
- **physical danger**: foreign bodies (piece of glass in a jar), metal, bone...
- **functional danger:** defects (of aspect, texture, deconditioning)...
- **administrative danger:** lack of labeling, abnormal delivery time...

The HACCP Approach in 13 Steps

HACCP action approach: 13 steps and 7 principles

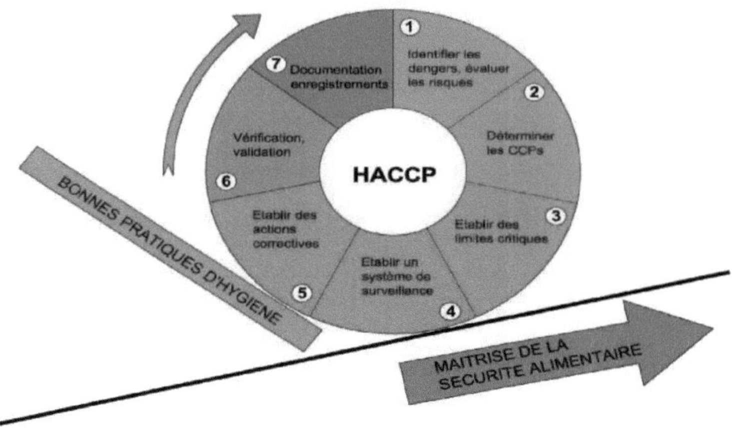

The HACCP method

Step 1- Define the scope of the study

A HACCP study applies to a single product (or a family of similar products from the same plant), for a single manufacturing process, against a group of identified hazards.

Step 2 - Set up the HACCP team

We gather a small multidisciplinary and competent team (5-7 people for large companies, 2-3 people for SMEs). Real and varied skills must be gathered: technical expertise is essential for steps 7-8-9. The team organizes itself (with a facilitator and a secretary), and is trained in the HACCP method. The team sets up a schedule with a deadline, establishes the tasks of each person and the deadlines. The team must have the necessary resources (time, money) and information

Step 3- Describe the product: a product audit.

For each component or product, precise data are collected: name, nature, form (volume, structure), % in the final product, preparation, treatments, storage conditions (duration, T°C), physical and chemical parameters that condition the development of bacteria or molds (aw, pH, contamination), distribution conditions. The product is described at the main stages of production (precise sheets), from the beginning to the end of the process.

- Input formulation = all raw materials, all ingredients.

- Describe in-process intermediates

- Formulation of the final product

Step 4- Identify the expected use of the product.

We examine the conditions of use at the factory, at the distributor (duration and temperature of conservation) and at the final users (consumers). Indeed, depending on the sensitivity of the consumer and the way the product is used, the same danger does not have the same consequences.

- Some users are especially sensitive

- Certain types of use reduce the risks

- But we must also consider "abnormal" conditions of use

Step 5- Make a manufacturing diagram: a process audit.

To make the diagram, the process is broken down into elementary operations (a simple diagram is made), noting for each step precise technical information, in particular their duration (but also the premises, the equipment, the sequences, the physico-chemical conditions such as temperature, pH and aw, the fluids or personnel, the contacts...). The interfaces are also described, for example transport and time between two operations.

Step 6- Check the manufacturing diagram, on site.

The HACCP team goes on site, on the production line, or in the kitchen, during operation, and checks that the diagram corresponds to reality. The steps below (7-8-9) are the most important. This is the "heart" of HACCP.

<u>**Step 7- *Hazard Analysis***</u>

A hazard threatens the safety of a person (*hazard= the potential to cause harm*). A danger is concrete (toxin, microbe, metal, chemical product...)
More broadly, a "hazard" is something that is harmful to the consumer.

A *risk* is the probability of the hazard occurring.

The 4 sub-steps of the hazard analysis are the study of Hazards, Risks, Causes and Preventive Measures.

The hazard analysis is done as a team, with everyone contributing their ideas and knowledge.

Identify Hazards

Health hazards are contamination by, or growth of, pathogenic bacteria in a food, the presence of toxins or chemical contaminants, ...
These hazards are identified by collecting published information, or collected from consumers (surveys, returns, complaints).

Evaluate the Risk of each hazard. Risk=Frequency x Severity

For each identified hazard, the risk for the product is evaluated: frequency and severity of the hazard (for the consumer, and for the company). This allows us to rank the hazards. For a given product, we focus on one (or a few) hazard(s), we leave the others aside.

Find the Causes

We place the hazards that "arrive" on the operations of the manufacturing diagram (done in steps 5 and 6). For each operation, we look for the causes of the hazards identified above. To find the causes of the microbiological hazards, we use the "5M".

Raw materials, equipment, working environment and methods, and especially the workforce are sources of microbiological hazards at every stage. The most important: the workforce, neither trained nor motivated, often "occasional".

Examples of 5 M:

Material: the milk that enters a cheese factory

Material: knives and tools, machines, packaging.

Environment: air, walls, carriers, tables.

Method: forward, refrigeration, waiting, recipe, cooking.

Workforce: hygiene training, cleanliness, (healthy) pathogen carrier.

<u>*Identify preventive measures for each operation*</u>

We move from the hazards and their causes to the identification of preventive measures, actions intended to eliminate the hazards, or to reduce them to an acceptable level. The preventive measures are often classic and are part of the good manufacturing practices (refrigerate, cook, train the staff. In ISO 22000 these are the "prerequisites"), or obvious (repair or change what is malfunctioning), but sometimes require creativity (change the process, reverse two steps, buy new equipment).

Step 8- Identify CCPs:

CCP = Point of Essential Control (PEC).

Let's explain it a little better: A CCP is a key step in the diagram, or an ingredient,

- Where a hazard can be controlled: possible control

- Where you MUST control the hazard: This step is essential.

A CCP is an operation whose non-control leads to an unacceptable risk, with no possibility of subsequent correction (*e.g. for a can, the closure is a CCP, because a can that is not tight is unacceptable, and cannot be "re-sealed" later*).

You can't monitor everything well. We therefore monitor the essential: the CCPs. Each CCP is a choice, a decision of the team: Sometimes we can decide to reduce the danger to an acceptable level (= to control it) at one stage rather than another. Another team could have made a different choice.

The team examines the diagram:

- Every raw material,

- Each intermediate product and the finished product,

- Each step, to see if it is a ccp:

To identify CCPs, we use the definition of a CCP and the CCP decision tree below:

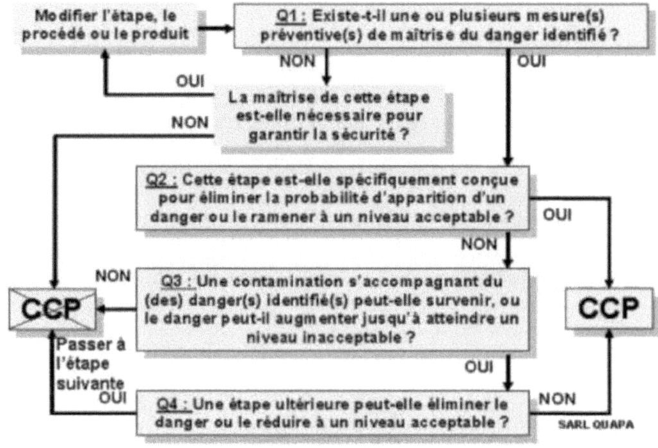

CCP decision tree

If so, this is the subject of the following HACCP steps (steps 9-10-11)

If not, the operation is not a CCP: it is a risky step, to be closely monitored. An essential point, yes. But no control.

Step 9- Establish critical limits & target levels for each identified CCP.

We want to prevent identified hazards. This is done by monitoring the CCPs. For each CCP, we look for the parameters that need to be monitored (*e.g. temperature, flow rate, duration, pH, concentration...*), and we decide on the critical limit that should not be exceeded, to ensure the control of the CCP. The critical limit is the numerical value that separates the acceptable from the unacceptable (safe product / dangerous product).

Step 10- Establish a CCP monitoring system

The team chooses, and describes on a form, the means used to monitor and control the CCPs and to ensure that the critical limits are not exceeded: automatic and continuous means are preferable (*e.g. thermal probe connected to a computer*). In practice, there is often discontinuous monitoring, the frequency of which must be specified

(*e.g. reading of T°C every hour*). The correct result is a release: if the critical limit is not exceeded, the product is released, it does not present a danger. It is also specified how it is done (procedure), who does it (responsible), and how the results are recorded (table, register, computer) to be able to "prove" the monitoring.

Step 11- Establish a corrective action plan: process and product.

Corrective action = what to do if critical limits are exceeded. After the prevention steps (9-10), correction steps are needed for each CCP: what to do if the results are not releasing? We specify in advance how to correct the process and the product:

1- How to get back to good working order (re-establish the control of the CCP at the process level) 2- What to do with the product (non-conforming product) These corrective actions must be described in a document, which specifies the person responsible.

The operator knows in advance what he has to do. He records when and how he applies the correction (nature and cause of the problem, quantity of product affected,...).

<u>Step 12- Establish documentation: plan, procedures, and records.</u>

The documentation has three components: plan, procedures and records.

1- the Haccp plan = the study itself and its verification (steps 1 to 13).

2- procedures = instructions for product compositions, flow chart operations, CCP monitoring systems and preventive (target) and corrective actions.

3- the recording of monitored values, manufacturing controls... signed by the operator. These records accumulate over time, and we must provide for their archiving (proof of the application of the HACCP plan, it is the demonstrative documentation).

Documentation is cumbersome to put in place (it replaces the oral tradition), but then saves time (to train a new employee, to respond to a customer request, or to allow an audit by a veterinary inspector).

<u>Step 13- Verify the system: compliance and efficiency</u>

Two aspects must be checked:

1/ that the system implemented in practice complies with the HACCP plan

2/ that this system is effective for safety. When the HACCP plan is put in place, it is planned how to verify compliance and effectiveness, and these verification provisions are written down. If the system is found to be ineffective, the HACCP study must be repeated.

Printed by Books on Demand GmbH, Norderstedt / Germany